Effective Risk management in the the UK construction industry

Abstract

The construction industry is inherently high risk due to the fact that all projects are unique, complex, and dynamic and involve significant levels of uncertainty. Whilst history shows there have been many successful projects delivered, there have also been numerous high-profile examples of failures which have been attributed to poor risk management.

Current research clearly establishes the benefits of effective risk management processes and how they can be used to best effect in particular situations. The literature however tends to focus on the advantages and disadvantages of particular aspects of risk management, rather than assess the relative importance of the different factors which may impact on its overall effectiveness. This book therefore explores the relative importance and inter-dependencies between the following factors.

- ➢ Systems and processes
- ➢ Human actions and behaviours
- ➢ Contract delivery and procurement strategy

In order to conduct a comprehensive investigation a qualitative review of the literature, published by recognised industry experts, was undertaken. Additionally, to obtain leading edge data within the industry, interviews and surveys were undertaken with a range of experienced practitioners, principally from contracting backgrounds. To deliver focus, data collection was centred on the aforementioned themes, with consideration given to the inter-dependencies between each of the factors.

Upon analysis of the data, significant evidence was found to support the view that formal risk management processes lead to significant benefits. However, it was also found that there is a clear need for simplicity and for experienced and capable people to take full advantage of these or to deviate from the formal processes where suitable. The primary research, in particular highlighted the significance of human actions and behaviours, in that interviewees felt their decisions were heavily influenced by those around them. This view was supported by a number of case studies which also highlighted the importance of using appropriate contract delivery strategies to ensure suitable risk allocation from the outset.

Overall it was found that, whilst appropriate systems and processes are important, as is the contract delivery strategy, the factor which ultimately determines success or failure is how people act and behave when implementing these. It was however noted that the actions and behaviours of individuals are heavily influenced by a number of factors, ranging from how they are motivated through to the capabilities of those around them and the suitability of the processes they are required to use.

In summary the research found that the key to improving the risk management effectiveness within the industry is improving the alignment of organisational and individual goals to encourage the most beneficial behaviours. In conjunction with this it is also vital that systems and processes are fit for purpose in terms of taking account of people's capabilities and ensuring that deficiencies are exposed and addressed before they have an impact.

Acknowledgements

I would like to thank my wife Ellie Cole for the moral support she has given me throughout and without her comfort and guidance, would have not been possible.

Contents

1.0 Introduction ... 6
  1.1 Problem Specification ... 6
    1.1.1 Problems which have given rise to the research ... 6
    1.1.2 Aim of the research ... 6
    1.1.3 Objectives of the research ... 6

2.0 Literature Review ... 7
  2.1 Introduction ... 7
  2.2 Systems and processes ... 7
    2.2.1 Introduction ... 7
    2.2.2 Description of a typical risk management process (RMP) ... 8
    2.2.3 Motives for formal risk management processes ... 8
    2.2.4 The benefits of simplicity versus of complexity ... 9
    2.2.5 The benefits of rigour versus flexibility ... 10
    2.2.6 The benefits of standalone processes versus processes which are fully integrated into project management ... 10
  2.3 The impact of humans actions and behaviours ... 11
    2.3.1 Introduction ... 11
    2.3.2 Leadership, organisational culture, skills and experience ... 12
    2.3.3 Communication and relationships ... 13
    2.3.4 Individual behaviours ... 13
    2.3.5 Conclusion ... 14
  2.4.0 Contract procurement and delivery strategy ... 14
    2.4.1 Introduction ... 14
    2.4.2 Responsibility for developing the strategy ... 15
    2.4.3 The factors affecting the choice of strategy ... 15
    2.4.4 Incentivising effective risk management through the delivery strategy ... 16
    2.4.5 Conclusions ... 18

3.0 Methodology ... 19
  3.1 Introduction ... 19
  3.2 Research methods and process ... 19
    3.2.1 Stage One: Desktop study (literature review) ... 19
    3.2.2 Stage Two: Semi - structured interviews ... 20
    3.2.3 Stage Three: Questionnaire survey ... 22
    3.2.4 Stage Four: Research analysis & conclusion ... 22

4.0 Primary research - outputs and commentary 23
4.1 Interview outputs 23
4.1.1 Interview responses on benefits of improving the effectiveness of risk management 23
4.1.2 Interview responses on comparing the importance of process, people and contract strategy 24
4.1.3 Interview responses on whether risk management is value adding or just a tool for improving forecasting accuracy 25
4.1.4 Interview responses on the impact of contract delivery strategy on the effectiveness of risk management 25
4.1.5 Interview responses on the benefits of simplicity versus complexity of risk management processes 26
4.1.6 Interview responses on the benefits of standalone processes versus risk management being an integrated part of project management 26
4.1.7 Interview responses on human influence and risk related behaviour 27
4.1.8 Interview responses on client influence on risk culture 28
4.1.9 Interview responses on appropriate risk allocation 29
4.1.10 Interview responses on effective risk management through incentivisation 29
4.1.11 Interview responses on the effectiveness of liquidated and ascertained damages 29
4.1.12 Interview responses on technological solutions to risk 30
4.2 Outputs from questionnaires 30
5.0 Limitations of the research findings 32
5.1 Primary research 32
5.1.1 Interviews 32
5.1.2 Questionnaires 32
5.1.3 Conclusion 33
5.2 Secondary research (literature review) 33
6.0 Primary and secondary research findings 33
6.1 Systems and processes 33
6.2 Human actions and behaviours 34
6.3 Contract procurement and delivery strategy 34
7.0 Conclusions 35
8.0 Recommendations 36
Bibliography 37
Appendices 41

1.0 Introduction

1.1 Problem Specification

1.1.1 Problems which have given rise to the research

The construction industry is inherently high risk due to the fact that all projects are unique, complex, and dynamic and involve significant levels of uncertainty. Project success is usually measured in terms of completion within time and budget, whilst also meeting the pre-requisite technical performance criteria. History shows that despite the successful delivery of many projects, both large and small, failure to stay within budget and meet deadlines is commonplace. Indeed, there have been a number of high profile cases of failure, for example, the Scottish Parliament building, where the final cost was more than ten times the original estimate (Potts, 2008).

Although some form of risk management process is usually used on larger projects, cost and time over-runs still occur frequently. Commonly, the extent to which risk management helps mitigate these issues is both difficult to assess and, in some cases, questionable.

There are a number of factors that can affect the success of risk management or otherwise, ranging from the suitability of the methods adopted, through to the influence of human behaviours on the implementation process. This research will consider the different approaches to risk management and its effectiveness across varying types of construction projects. It will also consider the factors that impact effectiveness and how improvements can be made in the future.

1.1.2 Aim of the research

The aim of this book is to identify the factors which lead to effective risk management and to identify opportunities for improvement.

1.1.3 Objectives of the research

The aim of this book is broken down into the following key objectives:

- ➢ To identify the key factors which affect the success or otherwise of risk management within in the infrastructure sector of the UK Construction Industry.
- ➢ To understand how different approaches and underlying influences impact on the effectiveness of risk management.
- ➢ To identify opportunities for improvement.

2.0 Literature Review

2.1 Introduction

The literature review explores a variety of views on the significance of the following elements on the effectiveness of risk management in the UK construction industry:

- Systems and processes
- Human actions and behaviours
- Contract procurement and delivery strategy

Whilst this review attempts to consider these as independent factors, there is clearly some level of inter-dependency and this is recognised within the study.

2.2 Systems and processes

2.2.1 Introduction

Risk management is the identification, assessment, and prioritisation of risks and is defined in ISO 31000 as 'the impact of uncertainty on objectives' followed by the economic coordination of resources to minimise, monitor, and control the probability or impact of unfortunate events. Equally, risk management is used to maximise the realisation of opportunities. It is therefore, by definition, a process which must be applied systematically in order to achieve the desired outcomes.

The Construction Industry Research and Information Association (CIRIA) guidance note on the Control of Risk, SP125 (1996) states that a systematic approach to risk management makes risks more explicit, formally describing them and making them easier to manage. It states that although you will never remove all the uncertainties, the approach does improve the chances of projects being completed on time, within budget and to the required quality. Chapman and Ward (2003) support this view and advocate a formal approach to risk management to ensure such benefits are achieved efficiently. They go on to suggest that proactive management of uncertainty leads to benefits well beyond control and neutralisation of threats.

2.2.2 Description of a typical risk management process (RMP)

The flow chart below, which has been taken from CIRIA guidance note SP125 (1996), illustrates a typical process for identifying and managing risks.

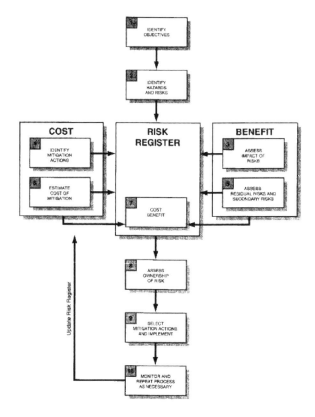

Figure 1 – Flow Chart on the typical process for identifying and managing risks (CIRIA Guidance Note SP125, 1996)

The process outlined essentially involves the identification of risks, followed by an assessment of the impact and the likelihood of them occurring. The risk register is used to capture the information relating to the status of the risk and to record ownership and risk mitigation responsibilities. Once commenced, the process described is a continuous and ongoing requirement to feedback and re-assess as circumstances change during the course of the project.

2.2.3 Motives for formal risk management processes

Chapman and Ward (2003) state that documentation is a key feature of all formal processes and they highlight that it is not only a key output, but it also facilitates the operation of the process. The underlying reasons for such benefits are considered under the following headings:

> ➢ **Clearer Thinking** – the process of having to write things down helps people think through the issues and results in a clearer overall understanding.

- **Clearer Communications** – Documentation provides the means of unambiguous communication at any point in time, this tends to result in fewer misunderstandings, particularly between people from different organisational units.
- **Familiarisation** – Documentation can provide the means of allowing new team members to become familiar with the issues quickly, this reduces the risks which can arise through miscommunication, especially where there is a high turnover of personnel on a project.
- **Record of Decisions** – Documentation can provide a record of the reasons behind certain decisions. This is important in some industries in terms of accountability of decision making, as a mechanism to learn from experience.
- **Knowledge Base** – Documentation can provide a means of capturing information that can be used to inform future decision making and risk management activities.

The view that some form of pre-defined process is fundamental to effective risk management is widely supported, however the literature review indicates differing views on importance of:

- Simplicity versus complexity
- Rigour versus flexibility
- Standalone processes versus processes which are fully integrated into project management function

The impact of these variables is considered under the relevant headings.

2.2.4 The benefits of simplicity versus of complexity

Chapman and Ward (2003) discuss the importance of ensuring risk management processes (RMP) are effective and efficient. They note that there is a cost to undertaking any risk management process to ensure the right cost-benefit balance is achieved. They suggest that in some circumstances, the process can be simplified, but emphasise that simplification can often result in reduced effectiveness and that it should never be undertaken to simply reduce cost.

Reducing the scope of analysis can be achieved in a number of different ways:

- The use of standard proforma, documentation and checklists
- Pre-specifying the form of qualitative and quantitative analysis to be undertaken
- Adopting single pass procedures that preclude revisiting earlier phases of the process
- Limiting the time and resources available for undertaking the risk management process

Resource availability is often a key driving force when determining a risk management approach. Depending on the resources available, it is sometimes necessary to prioritise risks and their impact as well as accept some level of risk in order to be effective. Accepting some level of risk by way of prioritisation of risk and its impact can be beneficial.

Restricting the focus of a RMP involves limiting the objectives that are set. One way of doing this, is to consider only significant threats to project performance, rather than all significant sources of certainty and their implications.

A common reason for RMPs being ineffective and inefficient, is lack of appreciation of the benefits of a comprehensive approach. This is often linked to a lack of organisational capability or investment in risk management. If comprehensive approaches aren't flexible, those responsible for RMPs will never learnt to appreciate the learning opportunities to manage risk more effectively. And if benefits are not appreciated, there will be limited investment in developing an organisational capability to obtain these benefits.

Organisations need an RMP which is both effective, in terms of the benefit for each project, and cost effective, in terms of the delivery of its effectiveness. To increase competitiveness within the industry, organisations ideally need to maximise the benefit of available resource invested in the time available. Rather than taking short cuts, which seek to cut the cost of risk management, the focus should be to apply risk management where the benefits are both obvious and significant. Additionally, organisations benefit from adopting efficient, streamlined processes designed for particular contexts.

2.2.5 The benefits of rigour versus flexibility

Whilst Chapman and Ward (2003) advocate a formal approach to risk management, they also recognise the benefit that flexibility can bring. They emphasise that if a flexible approach is adopted, it requires expert users to do this successfully and this is dependent on resource availability. A quality RMP should include specialists who are industry recognised experts with the knowledge and flexibility, and are able to appropriate their knowledge and experience. This results in making good judgements relevant to issues.

Chapman and Ward (2003) stress that whilst a flexible approach brings benefits, it also requires a cultural change and acceptance by the organisation. They state that "once people get used to making the distinction between action plans and other plans, they become more focused on exploiting opportunities associated with flexibility and on resisting costly changes in commitments."

Conversely, Smith (1999) states that risk management is a continuous and formal process and should span all the phases of the project. Risks and their impact should be observed on all the key sites of decision-making throughout the project, and by all participants in the decision-making process. In agreement with Smith (1999), Potts (2008) also identifies that risk management relies on a formal process. He states that the process should be headed by an experienced Risk Workshop Facilitator to ensure the RMP is administrated rigorously.

2.2.6 The benefits of standalone processes versus processes which are fully integrated into project management

Love and Li (1998) state that an integrated approach is required for risk management to be effective and that this should use Construction Process Re-Engineering (CPR). They consider CPR to be integrated and holistic in its approach. They highlight that its focus is on optimising process flow and eliminating waste whilst simultaneously fulfilling customer requirements, and satisfying the individual business needs of participating organisations.

Anumba and Evbuomwan (2000) developed an implementation model for process improvement for clients requiring processing in construction. They too, support a systematic approach to RMPs. In contrast, Sarshar et al. (1998; 2000); advocate SPICE (Structured Process Improvement for Construction Enterprises) which is a completely non prescriptive framework. SPICE describes the major process characteristics of an organisation at each maturity level, without prescribing the means for getting there. However, part of the SPICE methodology is to encourage a systematic approach to process improvement within construction through utilising lessons from other industries.

The level of formality in risk management can be considered a key factor in influencing an effective and efficient approach to risk management. Although a high level of formality usually will involve more cost, it will also result in more benefits. A less formal RMP involves less structure, documentation, clarification of objectives, deliverables, phases, and steps within a phase and fewer explicit phases. Informal risk management processes tend to result in a limited focus and thus are less effective (Chapman and Ward, 2003). Establishing the role of formality clarifies the need for a richer set of motives.

Incentive contracts motivate by specifying different sharing rates for costs above, below, and close to the target cost between the contractor and employer. This provides the opportunity to design a flexible incentive contract which can reflect the particular project context and unify motives based on the ability of the client and contractor to bear financial risk. Broome and Perry (2002) consider factors involved in the effective and efficient allocation of risk via incentive contracting. The underlying principle of incentive contracting being that the nature of the process aligns with the motivations of the parties involved to increase performance and more likely achieve project objectives. This is because the contractor and employer can negotiate responsibility of the constraints and risks based on the strengths and weaknesses of each party (Broome and Perry, 2002).

Incentive contracts and the flexibility they provide in terms of payment makes this RMP attractive to many organisations. However, the feasibility of widespread use is dependent on good working relationships, using obligational contracting rather than adversarial contracting.

2.3 The impact of humans actions and behaviours

2.3.1 Introduction

Despite advances in management training, risk management processes and the technology available to help us, there are still regular and numerous examples of unsuccessful projects. O'Neil (2014) considers that whilst failures are a result of inadequate project management application, RMPs should be able to expose issues and rectify or mitigate them to an acceptable extent and clearly this is not happening. Indeed, there is significant evidence to suggest that despite good RMPs, this is, at least in part, due to the effect of human behaviour.

From a different perspective, the Infrastructure Risk Group (2013) studied a number of major infrastructure projects on behalf of the UK government to identify best practices in relation to risk and contingency management. One of the key findings was that behavioural factors had a significant effect on project outcomes as well as the impact of risk management strategies and processes.

O'Neil (2014) identifies human dynamics within an organisation as the route to success or failure with regards to the effectiveness of risk management. He considers that human behaviour is largely set at a strategic level by the organisation culture or team. It is then influenced by team dynamics on a project level and further affected by the individual's actions, motivations and thoughts.

The impact of human actions and behaviours on the effectiveness of RMPs is therefore considered under the following headings:

- Leadership, organisational culture, skills and experience

- Communication and relationships
- Individual behaviours

2.3.2 Leadership, organisational culture, skills and experience

In order to understand the positive effect of creating the right culture, it is worth considering the background of two projects considered to be successes.

One such example was the design and construction of MediacityUK for the BBC. Former Lendlease Design Manager, Warburton (2014) noted that the establishment of a close knit team, comprising of the developer, the client and the contractor from an early stage, was key to managing the risks associated with a radical re-design and tight timescales. He also reinforced the importance of client leadership on complex projects, stating that the BBC should take credit for fostering the early collaborative approach to delivery.

The second example was the delivery of the London 2012 Olympic and Paralympic Park. Former Head of Projects for the Government Olympic Executive, Ian Williams (2014) largely attributes this success to having the right people with the right skills in key leadership positions. From the outset, the government appointed a specialist recruitment consultant to source key individuals with particular characteristics, these included having an understanding of how to achieve the right balance between cost and value, as well as having a resourceful and collaborative approach to overcoming difficulties. He also noted that having an immoveable deadline for completion, whilst posing significant challenges also led to a cultural benefit in terms of alignment of goals. Management of risk was central to the success with regular evaluation and adjustment of measures made to minimise or eliminate significant risks.

There is also considerable evidence to suggest that internal and external pressures can lead to organisational cultures which result in the inappropriate management of risks and contingencies (O'Neil, 2014). An example of this is where there is organisational pressure to win work in order to meet revenue targets. O'Neil (2014) suggests that this can lead to employees either consciously or sub-consciously producing an un-realistic budget and margin expectations for approval by senior managers. This can have a number of knock-on effects. One example occurs when contractors intend to use commercial means to maintain their profits, which subsequently drives risky behaviour, misreporting and a breakdown in team communication. In these situations, financial managers often exert pressure to reduce risk contingencies irrespective of actual risk levels to address more short-term financial needs. Tight margins, when winning work, often leads to a compromise in the quality or numbers of resources, which can result in a team lacking the necessary competence or motivation to perform and manage risk well.

Chapman and Ward (2003) also describe how culture can affect the management of risk. If an organisation does not distinguish between good luck and good management, there is the potential for managers to be considered wasteful if risk mitigation measures are seen to be unnecessary. This can lead to aggressive risk taking behaviours, which in the long run can have a detrimental effect.

The IRM report (2013) identified some unexpected outcomes which could be attributed to the maturity or culture of the organisation. One of the case studies considered is the London Underground's approach whereby project expenditure was authorised at a P50 level of risk. That is to say the authorised level of expenditure was based on the assumption that there was a 50% probability that sufficient risk moneys were allocated at the pre-construction stage to cover post-mitigated risk. The report states that this led to a number of unexpected behavioural issues, including the tendency for project managers to avoid developing early risk mitigation plans through fear of loss of contingency, which if later exceeded could have career limiting results. The report also identified that the approach led to a tendency for project managers to over-estimate the effectiveness of planned mitigations in response to cost pressures. Overall, it was noted that this type of optimism bias could lead to serious unexpected overspend when the mitigation measures were not as effective as expected.

London Underground overcame this by ring-fencing released contingency until the projects were complete (IRM report, 2013). The effect of this was to encourage managers to develop effective risk mitigation measures and to be transparent about their success without the fear of losing contingency which may be required for other unforeseen risks later in the project.

2.3.3 Communication and relationships

It is not surprising to find that organisations and teams who recognise the importance of leadership and culture also recognise the importance of relationships and communication in ensuring effective risk management. In the aforementioned examples of successful projects, there is clear evidence that strong relationships and good communication helped the teams manage risk and deal with issues effectively.

Warburton (2014) highlighted the benefits regarding the long term relationship between Lendlease and the BBC brought on the MediaCityUK project in terms of the alignment of values, as well as the technical understanding of requirements. He also emphasised that the good bond with fellow team members greatly enhanced team performance, as well as provided the platform to deal with difficult issues when they arose.

Williams (2014) recognised the benefits of weekly meetings between senior members of the various organisations involved in the delivery of the 2012 Olympic Park project. These were set up to deal with issues and make rapid decisions to address the evolving risk profile. He commented that not only did these result in solutions being quickly implemented, but they also led to excellent working relationships.

From another perspective, O'Neil (2014) states that whenever there is a problem, it can invariably be traced back to a breakdown in communication or reporting overriding the risk management controls in place. It is his view that this is a human problem even the best systems and processes cannot overcome. Such communication breakdowns can be accidental or deliberate and are most difficult to deal with when attributed to individual behaviours.

2.3.4 Individual behaviours

Individuals and their behaviours are an important aspect of effective risk management. When considering 'the team' or 'organisational behaviour' we are collectively considering a group of individuals and their behaviours in a group dynamic as well as how they act in isolation.

Smith, Myrna and Jobling (1999) suggest that there are two main categories of people: 'risk takers' and 'risk avoiders'. Whilst this is a simplistic statement which is not absolutely true, it is fair to say that some people are more likely to take risks than others, and that peoples' attitude to risk will depend upon their circumstances and personal experiences. They also suggest that 'risk takers' tend to under-rate risk and 'risk avoiders' see all the obstacles and tend to over-rate risk.

O'Neil (2014) discusses the fact that that as individuals we are also subject to bias, both on a conscious and subconscious level. Smith et al (1999) recognises the potential for bias and suggests that the impact of this should be minimised in risk workshops ensuring sufficient people are involved in the process and individual biases are counter-balanced. If this approach is taken, care is also needed in the team selection process so that there is a balance of different people, with a range of experiences. This ensures the team is not dominated by one or two people due to their nature or their position in the organisation.

Smith et al (1999) also discuss how the group approach can help avoid common pitfalls such as motivational bias (e.g. individual estimates reflecting the wishes of senior management) and the potential for individual experiences inappropriately affecting judgement. This can occur if people transfer knowledge and experiences from previous projects without recognising how the different circumstances affect the risk profile.

Many authors on the subject recognise the fact that individual behaviours can adversely impact the effectiveness of risk management, even when people are trying to do their best (O'Neil, 2014). We must however also recognise that not everyone works on projects for the good of the project and that sometimes people will deliberately do things for personal gain, thus undermining the RMP. O'Neil (2014) discusses various types of negative behaviour ranging from an unwillingness to recognise problems, through to ego-driven bullying. He also suggests that the ego or pride of individual managers often leads to the misguided intent to fix problems without telling others about them, only to realise too late that they cannot solve the problem alone.

2.3.5 Conclusion

According to Marchington and Vincent (2004), it is paramount that we distinguish between institutional, organisational and interpersonal forces that shape behaviour, in order to improve the effectiveness of risk management. We must therefore consider a strategy which encompasses process and human dynamics within the UK construction industry.

2.4.0 Contract procurement and delivery strategy

2.4.1 Introduction

The development and implementation of an appropriate contract procurement and delivery strategy is now widely considered to be one of the most important factors affecting risk management on major construction projects. Indeed, the strategy is the vital starting point for determining numerous key factors including risk allocation, types of risk management processes and stakeholder relationships.

The Latham Report (1994) and the Egan Report (1998) highlight the inefficiencies of traditional delivery strategies, particular noting the misconception that awarding contracts based on the lowest price bid, offers the best value. The Egan report (1998) suggests that savings of up to 30% could be achieved by the industry adopting a more collaborative approach, founded on a competitive process, with appropriate risk sharing. As a result of this, the UK industry has been looking to improve risk management through incentivisation schemes, improved collaboration and the integration of contract management with common goals and processes.

2.4.2 Responsibility for developing the strategy

According to Rowlinson and McDermott (1999), it is the client who is most affected by the outcome of a project, and this gives him the highest responsibility in terms of developing the risk management strategy. Langdon and Rawlinson (2006) build on this by saying all clients should devise their strategies based on project priorities, management capabilities and the extent of risk that they are prepared to accept.

Barnes (2006) encourages risk management through the development of contract strategies for complex projects. He also suggests that not all risks can be controlled by everyone; they should therefore be delegated to those best placed to deal with them. Griffiths (1989) further supports this view, stating that, "…the contract establishes the risk to be carried by each party. The general principle suggest[s]…that risks should be carried by the party best able to either control the risk or estimate the risk".

According to Aboushiwa & Bower (2000), "a contract strategy should be established at an early stage to meet the cost, programme and quality objectives of a particular project". The authors advocate the implementation of contract strategy as early as possible in a project, as it is the most important aspect of any project. This is because it forms a basis on which everything else in the project is built. Their view is that if a good contract strategy is implemented early, the activities are planned better and improved project outcomes will be achieved.

Whilst Venters (2005) is an advocate for the contract strategy to play a core role in success, he recognises that it is not the sole factor determining the success of a project, and that there are many issues which interdependently contribute to achieving success. He argues a "successful project must also require a solid work scope commitment of a project team and a realistic budget to support a successful project." In support of Venters (2005), Langdon & Rawlinson (2006) state that "it is the performance of the team that is at the root of project success not the strategy per se".

However, in support of Aboushiwa & Bower (2000), if an inappropriate strategy is selected, the inter-dependant factors which contribute to project success are critically affected (Smith, 1995). Thus, considering them as part of the contract strategy is a pre-requisite for achieving them (Smith 1995).

Alternately Perry (1985) states that the 'decision taken before the sanctions have the most significant impact'. He advocates that decisions made prior to the selection of the contract strategy, or even before the beginning of a project, can have the biggest impact on project outcomes. These can then limit or inhibit the availability of possible contract strategies available.

2.4.3 The factors affecting the choice of strategy

Bower (2003) considers contract strategy to be of utmost importance. Indeed, he states that "contract strategy has a major impact on the timescale and ultimate cost of the project". Equally, he considers that the negative impact on project outcomes through poor contractual strategic selection directly results in over budget projects and a delay in completion times. In support of Bower's view, Perry (1985) outlines the key decisions which influence the success for contract strategy, these are:

    a) Project characteristics

    b) Organisational structure

c) Contract form and type

d) Tendering process including conditions of contract, contract selection and tender analysis.

Perry (1985) also gives consideration to what makes an effective strategy. He explains that an "optimal contract strategy will be one which displays a consistent integration of the selection within each of these strategic areas". However, he notes that organisations can never know what the most effective strategy is, until the project has been completed and can be reviewed retrospectively.

Perry (1985) also highlights that whilst the client may have identified the best contract strategy at the time of its creation, in reality, it may not be the most feasible strategy to implement.
The author suggests there is no single best contract strategy, but selection must be based on being able to meet the objectives of the project. In support of this view, Bower (2003) states that "due to the diversity of both construction and the Client's requirements, there are different types of strategies available and no single uniform approach to contractual arrangements shall be advocated". The client therefore has the responsibility for not only selecting the contract form and type, but also deciding how to amend the standard forms to suit their requirements and ensure the most effective risk management strategy.

Smith (1995) suggests that the key areas which are considered when creating a contract strategy include systems and processes, risk allocation preferences and prioritisation of project objectives. Aboushiwa & Bower (2000) broadly support this view and stress the importance of pre-requisite activities in leading to a successful contract strategy selection. In opposition, Haswell & De Silva (1989) believe that the clients should focus primarily on contract type and contractor selection and that this is key in obtaining the optimum balance of cost and value. The table below summarises the authors' relative views.

Key Factors when Considering Contract Strategy	Authors		
	Smith (1995)	Aboushiwa & Bower (2000)	Haswell & De Silva (1989)
Project Objectives	✓	✓	✓
Organisational Systems and Processes in Relation to Design and Implementation	✓		
Risk Allocation	✓	✓	
Payment Terms	✓		
Contract Conditions	✓	✓	
Tendering Procedures	✓	✓	
Priorities		✓	
Responsibilities		✓	
Organisational Structure		✓	
Types of Contract		✓	✓
Contractor Selection		✓	✓

Table One: Comparison of authors views on key factors which should influence contract strategy

2.4.4 Incentivising effective risk management through the delivery strategy

Historically, clients and contractors have often tried to manage risk by its transfer onto other parties through negotiation. Olsen & Osmundsen (2005) suggest that "the Contractor can be motivated to control and possibly reduce construction costs by making him bear some risk". He then goes on to add that "trading off risk bearing and incentives, the buyer will offer more incentive based compensation (less cost sharing) in cases where the remaining project risk is low". The author suggests that financial incentives can increase risk management efficiency in projects, by underpinning factors which positively influence human motivation to achieve project goals. Incentivisation should be considered when creating a contract strategy. According to Bower & Merna (2002), there are three key types of financial incentivisation in construction contracts:

> 1. **Share saving incentives**: Also known as the "Win to Win Model" or "Guaranteed Maximum Pricing" (GMP). This is where cost savings on a target cost are shared between the Client and the Contractor based on an agreed formula in the contract.
>
> 2. **Schedule incentives:** This is where a bonus is offered to the Contractor for the early completion of a project; and
>
> 3. **Technical performance incentives:** If performance targets are met based on technical factors such as quality and functionality, a performance bonus is awarded to the Contractor.

Some studies (Kumarasaswamy and Dissanayaka, 2001; Hinze, 2002) have reported no significant effects, whilst others, (Ashley and Workman, 1986) despite reporting marginal improvement in performance quantity, found significantly more contractual disputes (A. Anvuur, 2010). Studies on cost incentives have generally produced mixed results. Bower and Merna (2002), reported better flexibility in planning and improved teamwork, but no clear effect on outturn project cost. Others reported tangible improvement on performance outcomes (Hauck et al.,2004).
Explanations for the inconsistent findings have often involved references to, amongst others, level of analysis confounding (Bresnen and Marshall, 2000) how even the most 'objective' key performance indicators are at best inadequate, can be highly corruptible (Fernie et al., 2006) and can cause incentive contingency misalignment (Hinze, 2002).

According to Bower et al. (2002), the fundamental principle of incentivisation is that "joint risks unify motives". This indicates the centrality of human motivation in an effort to align parties' goals with project objectives. According to field theory (Gold, 1999), motivation for human behaviour is a function of both the individual and environmental demands. Extrinsic factors stem from contractual obligations and formal control structures and link individuals' outcomes, in terms of rewards and sanctions, to some standard(s) of performance. On the other hand, the internal factors stem from an individual's desire to maintain and project a favourable social identity, which leads them to engage in discretionary and value-expressive behaviours.

Pattison (2012) advocates the use of the NEC suite of contracts (published by the Institution of Civil Engineers) to facilitate appropriate risk allocation as well as to help ensure good risk management practices during the course of project delivery.

Figure Two below illustrates the author's view on how differing options within this suite can be used to allocate risk between client and contractor.

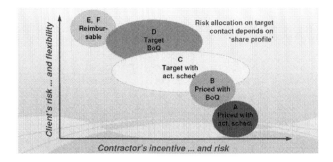

Figure Two – Options for Risk allocation in NEC Contracts Options (Pattison, 2012)

The options at either end of the spectrum of risk allocation, within the boundaries of the contract, are Options 'E' (which is fully cost reimbursable and where the majority of the risk is carried by the Client) and Option 'A' (which is effectively a lump sum contract, where the majority of risk is carried by the contractor).

Option 'C' in this form of contract enables an even balance of risk sharing and this is often used to jointly incentivise both parties to collaborate and manage and mitigate risk for their mutual benefit. Pattison (2012) emphasises that using the NEC suite of contracts does not provide an easy option and that regardless of the form of contract, there is still a need to follow a defined processes and to have a transparent approach to mitigating risk.

2.4.5 Conclusions

It is clear from the research that the choice of delivery strategy, which is often determined by the Client, can have a significant effect on how risk is managed and hence the effectiveness of the process. The literature suggests that the key factors in determining this are the timing of involvement for the various parties, and how the risk is allocated to each party. A number of authors emphasise the importance of allocating risk to those who are most able to influence the outcome. The research also highlights the impact strategy can have on individual and organisational behaviours and emphasises the importance of unifying motives at this stage.

3.0 Methodology

3.1 Introduction

Current research clearly establishes the motives for risk management and how particular approaches influence its effectiveness. As the literature tends to focus on specific areas of risk management, the methodology is designed to enable an assessment from a wider and more holistic perspective, thus enabling an understanding of the interdependencies between specific areas.

The research methodology was structured to enable exploration of emerging themes using a systematic and robust process. The literature review laid the foundations for the research and provided a benchmark to compare and contrast with the primary research, as well as dictating the themes to be explored. Primary research in the form of interviews for a range of industry practitioners was then conducted along these identified themes and compared with the secondary research.

After the interviews, questionnaires were used to gather further qualitative and quantitative information on the range of opinions and to enable further exploration of themes which had emerged earlier in the process.

At each stage, the research undertaken was justified and contextualised to demonstrate how the information contributes to the overall aim of the research. The outputs from each stage were then compared and contrasted and the limitations of the results were considered before conclusions were drawn.

3.2 Research methods and process

The research process is explained in more detail as follows:

- ➢ Stage One: Desktop Study (Literature Review)
- ➢ Stage Two: Semi - Structured Interviews
- ➢ Stage Three: Questionnaires
- ➢ Stage Four: Research Analysis & Conclusion

3.2.1 Stage One: Desktop study (literature review)

The desktop study was a structured analysis and critique of current findings from a range of academic and industry sources. The initial focus was to obtain a broad understanding of the range of risk management approaches across the industry. It was then followed by an assessment of the effectiveness of these approaches from the authors' perspectives and their views on how this could be improved.

Evidence was sought on how the authors assessed the impact of RMPs and how they have identified particular issues which have led to successful or unsuccessful outcomes. Once this was developed, it was used to provide a benchmark against which primary research was compared and contrasted.

The methodology utilised is summarised in the following flow chart:

Sources
- Academic
- Independent Bodies
- Private Industry
- Goverment Bodies

Taking account of:
- The range of qualifications and experience of the authors
- The coverage of the information published
- Gaps in research
- Relevance (dates / types of project)
- The authors views and recommendations

Identification of;
- Impacts of risk management
- Risk management practices and their effectiveness
- Issues in risk management and how they can be resolved
- Opportunities to improve the risk management process

Identification of;
- The extent to which risk management is effective
- Practices and underlying factors which are considered to result in successful risk management
- Practices and underlying factors which are considered to result in less successful risk management
- Potential improvement measures

3.2.2 Stage Two: Semi - structured interviews

Semi-structured interviews were undertaken with a range of industry practitioners. The semi-structured interview approach was used to enable the optimum balance between a standardised approach, and an open approach allowing the exploration of unexpected themes.

Using this approach, there was the potential for digression on to less-relevant subjects and the interview results could be more difficult to compare and contrast. This was managed by setting notional time limits for the response to each question, thus allowing the interviewer the opportunity to highlight the fact that responses were moving into areas which were not relevant to the research.

The main questions were prepared in advance, and this preparation included a number of supplementary questions which were designed to check understanding of the original question as well as to improve the flow of conversation should it stall. Care was taken to ensure questions were worded in such a way as to not lead interviewees, and to allow them to answer in a way that was relevant to them and their experience.

Whilst the research questions were ordered by theme, as this was expected to maximise the flow of conversation, interviewees were also encouraged to think holistically in their approach to answering and consider the inter-dependencies between the topics.

Prior to the interviews, the participants were given an information sheet detailing the nature and purpose of the research, together with consent forms enabling them to confirm whether they wished to proceed with the interview process. At this stage interviewees were made aware that, whilst their output would be summarised anonymously in the research, a dictaphone was to be used to record the interviews on the day.

Approximately ten participants from a range of professional backgrounds, experience and differing stakeholder interests took part in the interview process. Whilst this method allowed more opportunities to obtain a wider perspective, it was considered that this was a sensible and practical limit given the length of time it was likely to take to prepare and undertake the interviews, as well as complete the analysis. Care was therefore taken to ensure the interviewees had a reasonably wide range of experience and could provide sufficient insight into multiple industry backgrounds.

The overall aim of the interviews was to produce a rich set of data to compare and contrast with views spanning the industry as a whole, thus giving a wide and accurate representation as to what constitutes effective risk management. These views were then compared with that of each other and the themes that emerged from the literature review.

Interview Groups
- Academics
- Clients
- Contractors
- Client's Consultants

Preparation
- Identify key individuals with expertise
- Stakeholder Analysis
- Investigate interview techniques / methods
- Prepare interview questions

Conducting the Interviews
- Ask pre-defined questions (qualitative & quantitative)
- Note answers utilising a structured format
- Note othe indirect relevant information
- Explore unexpected areas of interest

Analysis & Conclusion
- Group and structure the information
- Analyse the results
- Summarise the output and draw conclusions

3.2.3 Stage Three: Questionnaire survey

The questionnaire incorporated a series of open and closed questions which were considered relevant to the full range of stakeholder groups, enabling both a qualitative and quantitative analysis. The aim of this was to enable a comparison of the output of the literature review, with the views of a range of practitioners involved in the delivery of projects and programmes of work.

The methodology aimed to ensure that the number and range of participants were adequate to give a fair representation and to minimise statistical bias.

Preparation
- Identify how the target audience will be reached
- How will the required number of surveys be obtained
- Identify questions based on the findings of the desktop study
- Structure the questions in a way which makes them accessible but also valuable to the research

Conducting the Survey
- Distribute the survey via professional networks, emails etc

Analysis
- Collate information
- Interpret results
- Analyse results and how they fit in with the research
- Understand the limitations of the results and potential bias
- Summarise Conclusions

3.2.4 Stage Four: Research analysis & conclusion

This stage involved collating the findings and results from the previous three stages and then undertaking a critical analysis of the information. The results from each stage were then compared and contrasted and an assessment on the effectiveness of the various approaches was made, taking into consideration different scenarios. Particular attention was given to identifying the underlying factors which contribute to successful risk management and understanding how this information could be used to improve overall performance.

4.0 Primary research - outputs and commentary

4.1 Interview outputs

Interviews were undertaken with seven people, whose roles and backgrounds are identified in the table below. Hereafter these will be referred to as Participant A, B, C, etc. as indicated below:

Role	Background	Reference
Commercial Manager	Contractor	Participant A
Project Manager- Client	Client	Participant B
Contracts Manager	Contractor	Participant C
Technical Project Manager	Contractor	Participant D
Senior QS	Contractor	Participant E
Associate Director	Client's Cost Consultant	Participant F
Regional Director	Contractor	Participant G

The comparative views of the interviewees are summarised below under the following headings:

- Benefits of improving risk management
- Comparing the importance of process, people and contract strategy
- Risk management - value adding or just a tool for improving forecasting accuracy?
- The impact of contract delivery strategy on the effectiveness of risk management
- Simplicity versus complexity of risk management processes
- Standalone risk management processes or an integrated part of project management?
- Human influence and risk related behaviour
- Client influence on risk culture
- Appropriate risk allocation
- Effective risk management through incentivisation
- The effectiveness of liquidated and ascertained damages
- Technological solutions to risk

4.1.1 Interview responses on benefits of improving the effectiveness of risk management

Risk management is often stigmatised as a tool which displays all good and bad news. It does however need to be perceived as a mechanism for improving overall performance through transparency. This can be achieved by reinforcing the material benefits of risk management (**Participant A**).

Risk management needs to be sold as beneficial and steer its users towards opportunities. When there is a low or no risk, there are always opportunities to improve the way in which project risks are managed with a greater focus on opportunity realisation. The benefits for opportunity are there, the industry and its people just need to realise it (**Participant F**).

One of the biggest challenges the industry needs to overcome, is how we unify goals and encourage collaboration on relationships which are short-lived, such as in the construction sector where projects are one-off and bespoke (**Participant F**).

The "you will act in mutual trust and cooperation" clause 10.1 of the NEC needs to be the heart of everyone's interest in setting the foundations of a good contract strategy; that should be the industry standard goal (**Participant F**).

4.1.2 Interview responses on comparing the importance of process, people and contract strategy

One of the key factors in effective risk management is the ability to identify and anticipate what may go wrong, the probability, the implications and impact. If people don't have the right skill set to do that, then often only the generic project risks are identified (**Participant C**).

Risk management is very much a human centric driven process and skills based application. Whilst projects need the right procedures, they also need the right people, with the right knowledge and experience, as risk management is a very fluid and complex subject matter. A project can have the right procedures and a group of people who don't fit the requirements for risk management, but it will limit the value of effective risk management. Conversely, the right people will make a bad system work **(Participant E)**.

In agreement, **Participant A** suggests that the processes are rigid and are similar across most of the industry, especially on large framework environments as the processes are well defined (**Participant A**). The processes aren't necessarily good or bad, what determines whether it works or not, is the people using it (**Participant A**).

The client or the contract will set out the process of managing risk (**Participant A**). The delivery strategy has the most effect, as it sets out how people work, what the processes are and how people interact with each other. If the strategy encourages people to work together to understand the real risks and come up with the best ways of dealing with them, then risk will be managed effectively and drive better project outcomes (**Participant G**). As an example, the National Grid uses large scale NEC frameworks which make risk management a central tool for its projects, integrating risk management as part of the construction process. This goes from target setting within the project, right through to closing it out. The risk register is a regularly used document and its use was encouraged at all the meetings, it formed a live document. Although the framework had a risk manager who administrated the processes, it formed part of the culture in running the framework. The Project Managers, the Quantity Surveyor and the Client, actively used the risk register to drive the project.

The same process was applied on other frameworks generally, as it's an NEC requirement. However, risk management is often seen as an administrative burden. The risks are ill thought about and once the risk processes are dealt with, are not used to manage a project. Even if processes and procedures are prevalent in the work place, people don't follow or us them properly and the project will not realise the benefit (**Participant A).**

Participant F concluded you can have a set of procedures which are robust and clear, without the right cultures and attitudes, people will always default to prior ways of working based on experience and judgement cues. To understand the right behaviours required, we have to look at delivery strategy and the pre-requisites for success in delivering that strategy (**Participant F**). Whilst organisations may have the right skills and attributes, they often don't have the resource due to budgetary constraints placed on them by the organisational policy. They therefore can't manage risk effectively to produce the best potential outcomes. We must consider the external constraints which may affect these (**Participant B**).

4.1.3 Interview responses on whether risk management is value adding or just a tool for improving forecasting accuracy

The intention for the risk management processes in the water industry is to reduce the outturn cost by mitigating risks. However, their intended effect is limited (**Participant G**). Risk management is used more to improve forecasting accuracy. The value is derived from the sense that if you don't undertake risk management or risk, costs are inevitably going to uncontrollably spiral. Risk management is beneficial in both respects, although it must be ensured that the focus is on risk management, not forecasting which is a side benefit (**Participant C**).

People in day-to-day decision-making, actively manage risk and reduce cost but formalising the risk management process does not have a large material impact. Businesses too often focus on risk registers and people lose focus on the key issues. If there was a more focused approach when looking at the key issues, people would take more ownership and do more to manage and mitigate significant risks (**Participant G**).

In contrast **Participant D** is of the opinion that there is a tendency to release risk monies to maintain forecast accuracy as opposed to improving accuracy. In contrast, it does encourage collaboration between the project team to consider the project risks. Project teams regularly update their risk register two to three times a month and this also acts as a mechanism to encourage communication and maintain a live document. If this process wasn't in place, the risk management meeting would probably not occur at all. It is also a tool which encourages project team discussions, and supports the allocation of responsibilities for managing risks (**Participant E**).

4.1.4 Interview responses on the impact of contract delivery strategy on the effectiveness of risk management

Even with a pro-active contract like the NEC, it requires a pro-active team to implement it and this can result in driving down costs and reducing programme delivery time, whilst setting the strategy for delivery. It is often wrongly used as a tool for driving target costs down prior to contract award in the hope that risks will be avoided or mitigated (**Participant A**).

Successful risk management drives value when it encourages a culture of risk mitigation. If risk is integrated as part of the incentive mechanism, then that's where risk management really adds value. The mechanism works by rewarding risk mitigation. If the contractor doesn't spend risk monies, then they are rewarded and share the benefit with the client if the risk remains unrealised. This incentive encourages the client and the contractor to jointly share and tackle risks (**Participant A**). On target cost contracts, the risk is not defined and is referred to as part of the target cost, this causes a lack of transparency when reporting risk with the client and vice versa. As the risk is an integrated part of the target cost, our client is only interested in whether he spends less than the target cost and does not consider risk to the extent that he should (**Participant A**).

From the contractor's point of view, they are then able to hide risk and spending by releasing funds. To overcome this behaviour, the culture has to be set by the client. The risk can only be unlocked by themselves, as they are the ones who can assist the contractor. For example, if the contractor has an unexpected risk with which only the client can help, i.e. the inability to get access to site, the contractor puts risk in the bid because the risk is not jointly discussed or inappropriately allocated. If there was an incentive mechanism for risk mitigation where the risks could openly be discussed, the contractor could offer savings to the client, mitigate the risks and reduce outturn costs. Otherwise self-interests and competitive behaviour takes a seat. (**Participant A**).

4.1.5 Interview responses on the benefits of simplicity versus complexity of risk management processes

If the culture is not there to drive effective risk management, it doesn't matter how complex the processes are, it won't work. However, the project team needs to appropriate the complexity of process not on the size of the project, but the level of risk. This also includes high risk projects which have a simple scope. A high risk project will still require a robust process to work, however, if it's too complicated, people won't follow it. This also includes looking at the skill base of the team (**Participant A**).

In contrast, the simpler the risk management process the better it is because people need to understand what they're doing and what impact it is having. As soon as it becomes too complicated people lose interest and ownership (**Participant G**).

Complex risk management processes may have their place on very large projects, but even then, the processes need to be broken down into manageable tasks and goals so those involved understand the impact they are having on the overall process (**Participant G**).

Participant C held the view that the model does not need to be complex, but the thought process involved in managing risk and managing risk effectively is a complex process. He states that the onus is on having a team with the right skills and attributes to identify and manage those risks.

4.1.6 Interview responses on the benefits of standalone processes versus risk management being an integrated part of project management

Risk management should be delegated amongst the project team members; however, the risk document should be managed by a central member. Technical backgrounds are often required to identify and communicate risks which allows the central controller to ascertain what the risks are, and if unrealised, what the remaining risk is. One person should own the document, but overall, it requires a team input (**Participant C**). In agreement, **Participant G** states that the risk management process should be integrated into project management because people with commercial roles often are the only individuals involved in the project who really understand the financial risks and opportunities.

Participant G however supports the use of a risk workshop, held by an experienced risk manager, who has an external view of the project because their level of independence encourages a broader, more holistic view. External facilitators have the ability to identify issues which people involved in projects are sometimes unable to, simply because of internal pressures. External facilitators are also not constrained by previous thinking (**Participant G**).

In comparison, **Participant A** states that whilst it may be appropriate to have a standalone risk management process led by a risk manager, when the process is too big to administer, there still has to be a team ethic regardless of whether there is a risk management team or not. The risk management team shouldn't exist to take ownership of risk. Their purpose is to administrate and implement the risk management process properly (**Participant A**).

If the contract and processes are set up properly, the individuals as well as the team should be rewarded for managing risk measured on the basis of project success. On National Grid projects, the fee was on mitigation. If the contractor saved one hundred thousand pounds for example, the client would give the contractor the fee on the saving even though the contractor hadn't spent it. Conversely, if the contractor spent half of that risk, they would only get half the fee. This resulted in management driving a message to the operations team that if they mitigated risks, they would make more money (**Participant A**).

4.1.7 Interview responses on human influence and risk related behaviour

From an organisational perspective, there can be a positive influence if risk is openly discussed. As for senior management, they have the potential to deliver the message that transparency will lead to better project outcomes. If the assessment of risk becomes burdensome, it often becomes a sole commercial function (**Participant A**).

However, pressure to win work is a prevailing attitude. When projects are negotiated, contractors often assess their own estimate against the client's budgets, and base their decision of acceptance on whether they can afford to meet the difference and compromise their assessment of risk. Therefore, there is a tendency to find ways of lowering the assessment of risk to meet client expectation, and therefore take higher risks on both of our behalf. This is an industry wide problem (**Participant A**).

In support of this, **Participant G** advocates a compromise in the assessment of risk and risk tolerance which is often dictated by market conditions. Organisations are much more likely to tolerate risk in harsh economic environments. If the project does not perform based on those expectations then pressures are exerted on people to be more optimistic in their reporting (**Participant G**).

Participant C, however attributes some of the influence to the nature of the contract itself which drives these behaviours. Due to defined cost mechanisms within some forms of contract, contractors have managed to compromise and negotiate a target price well below their original estimate. Because of the defined cost mechanisms, the contractor can sometimes accept a lower target cost and still make profit. This lack of transparency doesn't help.

Additionally, senior managers sometimes ask the project teams to make the project look like its performing better than it is, in order to bypass the risk management process and enter a contract unduly. Many issues occur when reporting is compromised and attempts are made to deliver unrealistic margin. To overcome this issue, contractor X changed the reporting standard so that risk would not form part of the margin, reporting it separately to increase transparency. Opportunity is now reported as a separate potential profit margin, but is no longer the focus. This means the risk remains in the control of the project team rather than senior management control. The expectation is now that contractor X can make the profit margins regardless of risk and opportunity (**Participant C**).

Participant F feels that there should be a greater emphasis on opportunity than risk. There is more value in opportunity as it is a proactive approach which saves money down the line. Trying to mitigate risk incurred means you're always looking behind (**Participant F**).

The focus needs to shift from cost to price. The price is where the biggest impact is and where disputes arise, which in themselves are costly. This would remove the risk from the delivery partner regarding their defined outturn cost. It also removes the risk from the client, as they equally can forecast their outturn more accurately. NEC Option C/D is an administratively burdensome contract. There is too much focus on cost, rather than the price and looking at agreeing prices ahead of time to encourage a rolling final account process (**Participant F**).

In contrast, **Participant G** believes the assessment of risk is down to individual behaviours and their need to protect their career interests and reputation. People hide contingencies within their budget because they don't want to be seen to fail if something unexpected happens. If people have a contingency fund to allow them to deal with problems without making them visible, their life is easier. When people view a project from the outside, if the outturn cost estimate fluctuates a lot, then the project is often seen to be out of control. Contingencies can be used to dampen the variances and this rightly or wrongly gives management more confidence (**Participant G**).

Retrospectively, project risks that could be mitigated are generally caused by individual behaviour. People try to deal with problems themselves too late rather than sharing them. The cause is a combination of people not wanting to be seen making mistakes or failing, as well as people's personal sense of pride and wanting to solve problems themselves.

To improve these behaviours a 'no blame' culture needs to be developed, allowing people to make mistakes and allowing them to include risk allowances for unknowns. These always happen, but there is often a reluctance to include contingencies for these things because people like to think that they have thought of everything (**Participant G**).

4.1.8 Interview responses on client influence on risk culture

The client is key to setting the culture and therefore delivering on the strategy. It's very hard for a contractor to set a positive behaviour if the client is not open to that agenda. This is often reflected in the contractual mechanisms and the amendments made to a contract. As soon as the client passes all the risks to contractor, gaming behaviours are encouraged, where both organisations are trying to benefit from each other's shortcomings, hiding information and misinforming one another. The contractor, often spends risk without bearing consequence because the client never knew about it in the first place (**Participant A**).

Additionally, the client often suffers from not having a fully defined scope, which means the contractor tends to build a lot of risk into the price. There needs to be more focus on having the right level of definition in the scope and on collaborating with the delivery partners to share ideas and get early involvement and input to narrow this gap (**Participant F**).

The clients should involve the whole team in setting the contract delivery strategy right from the beginning. This should start off with everybody getting a good understanding of the risks and opportunities and having a good think about who the best person is to manage them. There are some risks which should be managed by contractors because they have the most influence on risks like ground conditions, and there are other risks that are very much in the client domain (**Participant G**).

Looking historically at projects, there are a large percentage which are either over budget or in pain. At the tender stage, the client and contractor should sit down when writing the contract and clearly define what's included/excluded. Previously, the terms have been too vague and when things go wrong the relationship deteriorates and becomes adversarial (**Participant B**).

4.1.9 Interview responses on appropriate risk allocation

There has been a shift in the industry where the risk allocation has become more onerous on the contractor (**Participant C**). Clients can be too risk averse in areas where they are best placed to deal with it. For example, clients have been known to pass the risk onto the contractor for site access owned by the client. The contractor then has to price this risk, which costs the client additional money. It just doesn't make sense for a client which is clearly capable and has the resource to deal with that risk (**Participant F**).

If the project risk is appropriately allocated, and a contract strategy is developed which aligns everybody's goals and encourages them to work together, the best overall outcome will be achieved. The plan should involve everybody from the beginning and the strategy should encourage everybody to get a common understanding and work together towards the same goal (**Participant G**).

4.1.10 Interview responses on effective risk management through incentivisation

Contracts which use shared incentives result in better outcomes than those without, although the incentives need to be realised for all parts of the organisation. There are some clients who have shared incentive contracts using a target cost mechanism, but the contractual arrangement doesn't affect all key stakeholders in the client organisation to the same degree. Often it is the client's operational departments who can have a biggest effect on how risks are managed and mitigated, but they are not influenced by the shared incentive mechanism as they are funded from separate budgets (**Participant G**).

The issues arise when contractors are able to make profit well over the pain share mechanism due to defined costs. If there's a cap on the gain share, there is no incentive for the contractor to out-perform. There is often a clear benefit when going into pain in a target cost contract (**Participant F**).

Incentivisation promotes an interest and level of interdependency in tackling risks and succeeding in projects. The issue with a pain/gain mechanism is that when a project goes beyond the pain threshold, the client then loses the incentive to collaborate and tackle those risks. Capping gain can also limit the extent to which the contractor is incentivised to mitigate risk. However, there are incentive contracts where the target cost is fixed and there's no grace beyond, so if the project went over the client had no interest at all. So overall pain/gain incentives are a leading improvement in the industry (**Participant C**).

Alternatively, value can be driven by encouraging a culture of risk mitigation. If you make risk part of the incentive mechanism, for example, if you don't spend that risk you share the benefit of the unrealised risk, the client and the contractor can jointly share and tackle risks (**Participant A**).

4.1.11 Interview responses on the effectiveness of liquidated and ascertained damages

Liquidated and ascertained damages (LADs) aren't an effective incentive mechanism because they are often punitive and the cost of an overrun isn't significant enough in itself without having damages on top (**Participant G**). They are sometimes necessary from the client's perspective, but they should not be the sole focus of incentivisation as they often encourage unnecessary risk taking in the hope of avoiding the significant penalties (**Participant A**).

Positive incentives drive positive behaviour (**Participant A**). Using damages tends to lead to defensive behaviours and people pulling in different directions. However, sometimes there are situations where completion dates are important, or where there are large client costs if projects are not completed on time. LADs are a way of making sure that those vital completion dates are met. Normally they're not a good idea but there are some situations when they're probably the only option (**Participant G**).

Participant C however stresses that although they are not good for collaboration, they do make you think about risk early on in the project (**Participant C**). **Participant F** says that they are effective when they are relevant and used for the purpose of meeting regulatory targets. They must be justified. LADs are vital in a sector driven by time. They should be applied where there is a defined reason and a material loss incurred by the client. Using LADs as a penalty is never effective, it acts in the same way as penalising individuals, and it doesn't work. Clients should make provision for LADs where appropriate, but the focus on motivating the contractor to complete the project on time should be leveraged through good working relationships and assistance (**Participant F**).

4.1.12 Interview responses on technological solutions to risk

There needs to be two big shifts in the construction industry. One is a shift in behaviours towards a more collaborative approach and appropriate allocation of risk. The second is improvement in information and communication capabilities through BIM which will centralise and network relevant information (**Participant F**). **Participant B**, however highlights that for these technological improvements to be effective within the industry, it requires the buy in of those using it. If people don't take the time to provide accurate information of live and historic projects, then the outputs produced by collaborative technologies to improve the effectiveness of risk management will be constrained.

4.2 Outputs from questionnaires

Participant's views were obtained by assessing their level of agreement or otherwise with twelve unambiguous statements about the benefits or otherwise of alternative approaches to risk management. Respondents indicated their level of agreement with each of these statements by selecting a representative point on the linear scale as shown in Figure Three below:

Figure Three: Questionnaire Linear Response Scale

Prior to answering the questions on risk management, respondents were asked to confirm their background and level of experience. They responded as follows:

> Of the twenty-eight respondents. 60.7% were contractors, 28.6% were designers and the remaining 7.1% were from a client background (**Appendix A**).

- 64.3% of the participants had over ten years' experience in the construction industry and 25% between 5 and 10 years' experience (**Appendix B**).

Responses to the questions on the effectiveness of alternative approaches to risk management and other related matters are summarised as follows:

- 82% of the participants were of the opinion that in their working environment, the defined risk management processes and procedures are easy to understand and implement, of which 39.3% agreed on 6 out of 8 on the matrix rating scale. (**Appendix C**)

- Interestingly, 92% found that risk management processes have a significant impact on the success of projects with 40% agreed on a score of 7/8. Only 7.1% disagreed that risk management has a significant impact on the success of projects of which 100% of these participants chose 3 out of 8 on the scale (**Appendix D**).

- However, 35.7% (6/8 rating) of the participants were of the opinion that it is rare for a low risk, higher cost solution to be adopted when there is a lower cost alternative in its place. Overall 89.2% of the participants agreed to varying extents (**Appendix E**). In addition to this, the majority of participants believed that people are overly optimistic when assessing the effectiveness of the risk mitigation measures. 32.1% making the majority chose 6 out of 8 whilst only a total of 21.3% in total disagreed (**Appendix F**).

- Over 90% agreed collaborative contract with shared incentives for effective risk management leads to better outcomes with clients. In contrast only 7% disagreed that they would improve the performance of outcomes (**Appendix G**).

- On average, 87.8% made the conclusion that the main risks identified at the beginning of projects are usually the ones that come to fruition. Interestingly, the 32.1% felt that the risks identified, are not the ones which usually come to fruition (**Appendix H**).

- Everyone agreed (100%) to a varying extent that risk management processes should be an integral part of project management as opposed to having a standalone team. 46.4% chose 8 out of 8 whilst 39.3% 7 out of 8 (**Appendix I**).

- 57% believed the evaluation of risk and opportunity is influenced more than it should be by the expectations of stakeholder. 6 and 7 out of 8 both achieved a balanced score of 25% and 4 and 5 out of 8 both scored 17.9% (**Appendix J**).

- 89.4% were of the opinion that contractual arrangements and appropriate allocation of risk has a significant effect on whether a project is successful. Only 10.7% believed it did not (**Appendix K**).

- 81.2% came to the conclusion that risk is often passed onto parties who are lower in the procurement hierarchy whilst 17.8% did not (**Appendix L**).

- 38.5% jointly at 6 and 7 out of 8 felt that people did often avoid releasing unspent risk monies because of the fear of additional unforeseen risks whilst 11.5% in total to varying extent disagreed that this does occur often (**Appendix M**).

- ➢ 89.3% felt that contractual arrangements sometimes hinder a collaborative approach to mitigating risks compared to 10.7% who disagreed that this was the case (**Appendix N**).

- ➢ 79.6% disagreed that the key to improving risk management is in developing more complex processes to cater for increasing project complexity whilst 21.4% agreed that it would (**Appendix O**).

The questionnaire also included an option for respondents to insert comments describing how they believed risk management could be improved in the industry. Responses in this section were as follows:

- ➢ Risk management it needs to be seen as an overall project management tool and not a stand-alone commercial exercise done to affect the forecast with sensible and achievable mitigation plans and owners. Supply chain need to fully understand they play a role in risk management of any project they are involved in no matter how insignificant it may seem (**Appendix P**).

- ➢ The key to effective risk management is understanding the blockers early enough to do something about them. Effective communication within a team helps to achieve this (**Appendix P**).

5.0 Limitations of the research findings

5.1 Primary research

5.1.1 Interviews

Whilst the interviewees had a range of experience and professional backgrounds in the construction industry, all participants were at the time working in a framework environment with one particular water industry client. Six out of the eight were contractor employees, with one participant employed by the cost consultant and one employed by the client.

Even though the eight interviewees appeared to have considerable experience in risk management within their particular sector, their views cannot be considered representative of the whole of the UK construction industry.

5.1.2 Questionnaires

Whilst the number and range of people who completed the questionnaire were more than three times the number of people who participated in the interviews, the outputs cannot be considered to be representative of the whole industry.

It is however worth noting that the respondents' backgrounds were different from those of the interviewees. Whilst almost two thirds of the participants were employed by a contractor, and almost a third were employed by a design consultant, approximately one quarter of respondents were working for other clients within the industry.

On the basis that the number of respondents were low from a statistical perspective, and the respondents only represented a small part of the overall industry, little weighting can be given to the outputs.

5.1.3 Conclusion

Notwithstanding the question of the validity of the primary research, it is interesting to note that the views of respondents broadly aligned and supported the findings of the literature review. The interviews in particular highlighted a number of good examples of the benefits, or otherwise, of particular approaches and gave a good insight into the influence of human dynamics.

5.2 Secondary research (literature review)

Some of the literature reviewed was over ten years old, and whilst they may still be highly relevant, they may not be the most up-to-date sources of information. Additionally, some of the risk management theory is non-contextual and unrelated to construction risk management. The academic papers used may not have the experience to draw accurate conclusions and as a result, may have drawn from a series of conclusions made by others. Finally, the private publications used may contain bias due to vested commercial and public relation interests.

6.0 Primary and secondary research findings

The key findings of the research are summarised against the headings used in the literature review:

6.1 Systems and processes

Formal risk management processes and the associated documentation lead to significant benefits including clearer thinking, clearer communications, clearer definition of roles and responsibilities and better capture of knowledge for future use (Chapman and Ward, 2003). 82% of questionnaire participants stated that the defined RMPs and procedures were easy to understand and therefore easier to implement (Appendix C). The level of formality requires should not be dictated by the size of the project, but the level of risk. There is a greater need for a formal and robust process where the risk is greater, however, if the process is too formal and complex, people will not follow it (Participant A).

Under most circumstances risk management should be an integral part of the project delivery and should involve a wide range of expertise rather than being undertaken by the commercial team or by specialist risk managers (Love and Li, 1998). RMPs should be integrated into project management because those with commercial roles are often the only individuals involved who truly understand the financial risks and opportunities (Participant G). However, the risk document should be managed by a central member of the team. A technical background is required to identify and communicate risks which allows the central controller to ascertain what the risks are, and if unrealised, what the remaining risk is (Participant C). In conjunction with this, 100% of the questionnaire participants agreed that the RMP should be an integral part of project management as opposed to having a standalone team (Appendix I).

It is essential when determining the right level of input and the complexity of RMPs, that an accurate account is taken of the cost versus the benefit, as well as the capabilities of the people who are going to implement the process (Chapman and Ward, 2003). If the culture is not there to drive effective risk management, it doesn't matter how good the processes are, they will not be effective (Participant A).

The cost benefit balance and the abilities of those involved should be at the heart of determining the complexity of RMPs for particular circumstances. Simplification often results in better engagement due to the reduced administrative burden, but a complex project with a high level of risk may benefit from more complex processes (Participant A). 91% of participants in the questionnaire agreed that risk management procedures have a significant impact on the success of projects but to varying degrees (Appendix D). However, 79.6% disagreed that the key to improving risk management is in developing more complex processes to cater for increasing project complexity (Appendix O).

Although formal processes form a good basis for undertaking risk management, there is a clear need for experienced people who are able to deviate from the formal processes where appropriate. (Participant C).

6.2 Human actions and behaviours

There is significant evidence that the root cause of project failures which are attributable to poor risk management is linked to the way that people have acted and behaved rather than failings in the systems and processes (O'Neil, 2014). 57% of those surveyed, believe that the evaluation of risk and opportunity is influenced more than it should by the expectations of stakeholders (Appendix J).

People are central to the effectiveness of risk management because regardless of the process, it is ultimately people who have to make decisions on what the key risks are, how issues are prioritised and what the response will be to particular situations (Infrastructure Risk Group, 2013).

Human behaviour and interactions are very complex and outcomes are difficult to manage or predict. It is widely recognised that individual behaviours can adversely impact the effectiveness of risk management, even when people are trying to do their best. It is also often noted that everyone does not always act in the best interests of the projects that they are involved in, and this can occur as a result of issues ranging from personal pride through to deliberately malicious behaviours (Smith, Myrna and Jobling 1999).

There is significant evidence that the organisational culture and project environment have a large effect on how people act and behave. Evidence suggests that developing a non-adversarial and transparent approach where individual's goals are aligned with organisational goals, tends to lead to behaviours which support better risk management (Olsen and Osmunden, 2005). However, 89.3% of participants felt that contractual arrangements sometimes hinder a collaborative approach to mitigating risks compared to 10.7% who disagreed that this was the case (Appendix N). Contracts with shared incentives result in better outcomes than those without, although the incentives need to be realised for all parts of the organisation (Participant G).

6.3 Contract procurement and delivery strategy

It is widely accepted that the selection and implementation of an appropriate contract procurement and delivery strategy has significant impact on the effectiveness of risk management (Aboushiwa and Bower, 2000). This is generally attributed to the combination of ensuring appropriate risk allocation from the outset and creating an environment that encourages transparency and alignment of goals (Smith, 1995).

Whilst there is some evidence that the wholesale transfer of risk to design and build contractors has led to a number of highly successful projects, there is significant evidence to suggest that risks should be allocated to organisations which are best placed to manage them (Pattison, 2012). Primary research findings show that risks are often passed down the supply chain without considering whether those organisations are able to manage and mitigate those risks (Participant F). 81.2% came to the conclusion that risk is often passed onto parties who are lower in the procurement hierarchy whilst 17.8% did not (Appendix L).

Research indicated that since the publication of the Latham and Egan reports in 1994 and 1998 respectively, there has been a UK government drive to encourage the use of the NEC suite of contracts published by the Institution of Civil Engineers. Whilst this suite of contacts enables a range of risk allocation options, its use has led to improved clarity in terms roles and responsibilities for managing risk (Pattison, 2012). The use of shared risk arrangements between clients and contractors has also led to a more open and transparent approach than found in more traditional contracts (Bower and Merna, 2002).

7.0 Conclusions

The primary objective of this research was to identify and understand the key factors that influence the success or otherwise, of risk management across the infrastructure sector of the UK Construction Industry.

A number of key areas were found to significantly influence the effectiveness of risk management. These include ensuring appropriate risk management systems and processes are utilised, taking account of the impact of human actions and behaviours, and ensuring the procurement and delivery strategy aligns the parties to achieve the project goals. It was found that none of these factors stood alone and that the approach to each affected the level of influence that the others had on overall success.

There is significant evidence to support the view that formal risk management processes lead to significant benefits, however it was also found that there is a clear need for experienced people to take full advantage of these or to deviate from the formal processes where appropriate. The research highlighted the significance of human actions and behaviours and how decisions in relation to risk management are often heavily influenced by factors relating to personal circumstances or organisational culture.

On this basis, it is considered that whilst people's actions and behaviours are in many ways the key, other factors significantly impact these and strategies and processes to be designed to encourage the right behaviours, as well as act as a safety net, should people act inappropriately.

It is therefore considered that the key to improving the effectiveness of risk management in the industry, is improving the alignment of organisational and individual goals to encourage the most beneficial behaviours, whilst also ensuring that systems and processes are designed to highlight and address deficiencies before they have an impact.

8.0 Recommendations

Looking forward, it is recommended that further research is undertaken into the influence of human actions and behaviours on the success of risk management and in particular how individual and organisational goals can be better aligned to achieve the desired results.

As well as undertaking additional research, it is recommended that the industry invests more in developing delivery strategies and processes that encourage the right type of behaviours, whilst also ensuring sufficient visibility and control to minimise the risk of errors of judgment or inappropriate behaviours.

It is considered that the combination of these will lead to more visibility of issues and an increasingly proactive approach to risk management. This will ultimately result in a significant improvement in the effectiveness of risk management across the infrastructure sector of the construction industry.

Bibliography

Aboushiwa, M.A. & Bower, D. (2000) *Promoter Briefing on the Procurement and Selection of Contract Strategy Offering the Best Value for Money*. London: UMIST

Anvuur, A., Kumaraswamy, M. (2010) *Promises, Pitfalls, and Shortfalls of the Guaranteed Maximum Price (GMP) Approach: A Comparative Study*. London. University College London and University of Hong Kong

Ashley, D.B. and Workman, B.W. (1986) *Incentives in construction contracts*, Austin, TX: Construction Industry Institute (CII)

Assaf, A. & Al-Hejji, S. (2006) Causes of Delay in Large Construction Projects. *International Journal of Project Management.*, 24, 4, 349-57

Barnes, M. (2006) *Project Management in Uncertainty: An Integrated Approach*. Speech. Shanghai: 20th IPMA World Congress of Project Management

Boukendour, S. and Bah, R. (2001) The guaranteed maximum price contract as call option. *Construction Management Economics.*, 19, 6, 563-567

Bower, D. & Merna, A. (2002) Finding the Optimal Contractual Arrangement for Projects on Process Job Sites. *Journal of Management in Engineering.*, 18, 1, 17-20

Bower, D. (2003) *Management of Procurement*. London: Thomas Telford

Bower, D., Ashby, G., Gerald, K. and Smyk, W. (2002) Incentive Mechanisms for Project Success, 2002. *Journal of Management in Engineering.*, 18, 1, 37-43.

Bresnen, M. and Marshall, N. (2000) Partnering in construction: a critical review of issues, problems and dilemmas. *Construction Manage Economics.*, 18, 2, 229-237

Broome, J. and Perry, J. (2000) How Practitioners Set Share Fractions in Target Cost Contracts. *International Journal of Project Management.*, 20, 59–66

Broome, J.C. & Hayes, R.W. (1997) A Comparison of the Clarity of Traditional Construction Contracts and of the New Engineering Contract. *International Journal of Project Management.*, 15, 4, 255-61

Broome, J.C. (1997) *Best Practice with the New Engineering Contract*. Proceedings of the Institution of Civil Engineers., 120, 2, 74-81

Carty, G.J. (1995) Construction. *Journal of Construction Engineering Management.*, 121, 3, 319-328

Cerić, A. (2003) *Process Driven Risk Management*. Salford. University of Salford, 2003

Chan, D.W.M., Chan, A.P.C., Lam, P.T.I., Lam, E.W.M. and Wong, J.M.W. (2007) Evaluating Guaranteed Maximum Price and Target Cost Contracting Strategies in the Hong Kong Construction industry. *Journal of Financial Management Property Construction*, 12, 3, 139-150

Chapman, C. & Ward, S. (2003) *Project Risk Management*. Estados Unidos: Wiley

Dhanushkodi, U. (2012) *Contract Strategy for Construction Projects*. Manchester. Manchester University

Egan, J. (1998) *Rethinking Construction: The Report of the Construction Task Force*. London. The Department of Trade & Industry

Fernie, S., Leiringer, R. and Thorpe, T. (2006) Change in construction: a critical perspective. *Building Research and Information*, 34, 2, 91-103

Finnemore, M., Sarshar, M. (2000) *Linking Construction Process Improvement to Business Benefit*, Bizarre Fruit National Conference, University of Salford, 94-105.

Finnemore, M., Sarshar, M., Haigh, R. (2000) *Case Studies in Construction Process Improvement*, ARCOM Doctoral Workshop on Construction Process, Research, Loughborough University

Finnemore, M., Sarshar, M., Haigh, R., Hutchinson, A., Timms, R. (2000) *SPICE: A Structured Process Improvement Tool for Construction*, CIB TG36 International Conference on Implementation of Construction Quality and Related Systems, Lisbon

Gidado, K.I. (1996) Project complexity: the focal point of construction production planning. *Construction Management Economics*, 14, 3, 213-225

Godfrey, P. (1996) *A Guide to Systematic Management of Risk from Construction*. London. CIRIA

Griffiths, F., (1989). Project Contract Strategy for 1992 and beyond. *International Journal of Project Management*, 7, 2, 69-83

Haswell, C.K. & De Silva, D.S. (1989). *Civil Engineering Contracts: Practice and Procedure*. 2nd edition. London: Butterworths

Hauck, A.J., Walker, D.H.T., Hampson, K.D. and Peters, R.J. (2004) Project alliancing at National Museum of Australia - collaborative process. *Journal of Construction Engineering Management*, 130, 1, 143-152

Hinze, J. (2002) Safety incentives: do they reduce injuries? *Practical Periodical on Structural Design and Construction*, 7, 2, 81-84

Hughes, W., Yohannes, I., Hillig, J. (2007) *Incentives in Construction Contracts: Should we Pay for Performance Contracts*. CIB World Building Congress

Infrastructure Risk Group (2013) *Managing Cost Risk & Uncertainty in Infrastructure Projects*. The Institute of Risk Management

Jackson, S. (2002) *Project Cost Overruns and Risk Management*. Reading. University of Reading

Kalvet, & Lember, V. (2010) Risk management in public procurement for innovation: the case of Nordic Baltic Sea cities. *Innovation - The European Journal of Social Science Research*, 23, 3

Kamara, J.M., Anumba, C.J., Evbuomwan, N.F.O. (2000) Process Model for Client Requirements processing in construction, *Business Process and Management*, 6, 3, 251-279.

Kumaraswamy, M.M. and Dissanayaka, S.M. (2001) Developing a decision support system for building project procurement. *Building Environment*, 36, 3, 337-349

Kwawu, W., Laryea, S. (2013) *Incentive Contracting in Construction*. Reading. University of Reading

Lampman, R.J. and Dimeo, B.S. (1989) Team collaboration like playing ball. *Journal of Real Estate Development*, 5, 1, 56-62

Langdon, D. & Rawlinson, S. (2016) *Procurement: Construction Management,* 2006. [Online] Available at: <http://www.building.co.uk/data/procurement-construction-management/3072705.article# > [Accessed 4th April 2016].

Latham, M. (1994) *Joint Review of Procurement and Contractual Arrangements in the United Kingdom Construction Industry.* London. Construction Excellence

Lewis, H., Allan, N., Ellinas, C., Godfrey, P. (2014) *Engaging with Risk.* London. CIRIA

Lock., D. (2007) *Project Management: Project and Programme Management Resources for Students.* 9th edition, Gower

Love, P.E.D. and Li, H. (1998) From BPR to CPR - Conceptualising Re-engineering in Construction, *Business Process Management,* 4, 4, 291-305

Marchington, M., & Vincent, S. (2004) Analysing the Influence of Institutional, Organizational and Interpersonal Forces in Shaping Inter-Organizational Relations. *Journal of Management Studies*

Masterman, W.E., (1992) *An Introduction to Building Procurement Systems.* 1st edition. London: E & FN Spon.

National Audit Office. (2010) *PFI in Housing.* London. National Audit Office

O Neil, C., Hansford, P., Gaitskell, R., Horne, R., McArthur, J., Messenger, J., Moore, E., Somerset, D., Thomson, G., Stephen, Warburton., Watson, A., Whitson, G., Williams, A. (2014) *Human Dynamics in Construction Risk Management.* London. Chartered Institute of Arbitrators

Olsen, T.E. & Osmundsen. (2005) Sharing of endogenous risk in construction. *Journal of Economic Behavior & Organization,* 58, 4, 511–526

Oughton, J., Omand, D. (2004) *Managing Risks with Delivery Partners.* London. Office of Government Commerce

Patterson, R. (2012) *Using NEC Contracts to Manage Risk and Avoid Disputes.* Surrey. Mott Macdonald

Perry, J.G. (1985) *The Development of Contract Strategies for Construction Projects.* Ph.D. thesis

Potts, K.F. (2008) *Construction Cost Management.* London. Taylor & Francis

Pryke, S. and Pearson, S. (2006) Project governance: case studies on financial incentives. *Building Research Information,* 34, 6, 534-545

RICS (2007) *Contracts in use: a survey of building contracts in use during 2007.* London: The Royal Institution of Chartered Surveyors

Rowlinon, S. & Mcdermott, P. (1999) *Procurement Systems: A guide to best practice in construction.* 1st edition. London: E & FN SON

Saminu, S., Prasad, R., Thamilarasu, V. (2015) A Study of Various Factors Affecting Risk Management Techniques in Construction Projects: A Case Study of India. Bangalore. *International Journal of Research and Engineering Technology*

Sarshar, M., (1998), *Standardised Process Improvement for Construction Enterprises (SPICE).* Proceedings of 2nd European Conference on Product and Process Modeling, Watford

Smith, J. (1995) *Engineering Project Management.* 2nd edition. London. Blackwell Science

Smith, J., Merna, T. & Jobling. (2006) *Managing Risk in Construction Projects*. 2nd edition. London: Blackwell

Smith, N., Merna, T., Jobling, P. (2006) *Managing Risk in Construction Projects*. Oxford: Blackwell Publishing

Smith, N.J. (1999) *Managing Risk in Construction Projects*, Blackwell Science

Stukhart, G. (1984) Contractual Incentives. *Journal of Construction Engineering Management*, 110, 1, 34-42

The Tavistock Institute. (2000) *Setting Incentives and Shared Saving Schemes*. London. The Tavistock Institute

Turner, J.R. (2004) Farsighted project contract management: incomplete in its entirety. *Construction Management Economics*, 22, 1, 75-83

Venters, S. (2005) *Importance of Contract Strategy - A case study of the OLE and distribution Alliance project*,

Zartman, I.W. (2005) Comparative case studies. *International Negotiations*, 10, 3-16

Appendices

Appendix A

Your background in the industry (28 responses)

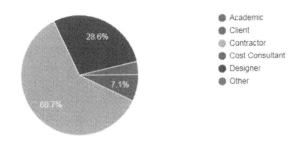

Appendix B

Length of Experience in the Industry (28 responses)

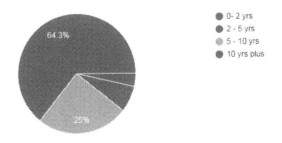

Appendix C

In my working environment the defined risk management processes and procedures are easy to understand and implement
(28 responses)

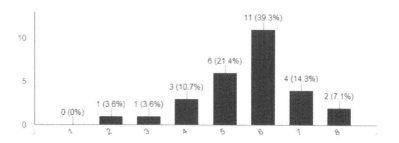

Appendix D

In my working environment the use of risk management processes and procedures has a significant impact on the success of projects
(28 responses)

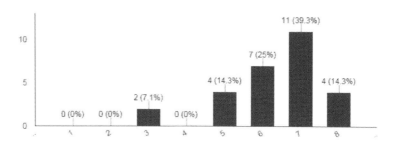

Appendix E

It is rare for a low risk, higher cost solution to be adopted when there is a lower cost, but higher risk alternative which appears viable
(28 responses)

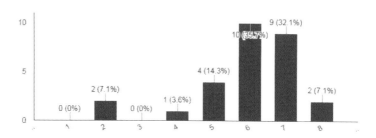

Appendix F

People are overly optimistic when assessing the effectiveness of the risk mitigation measures that they are planning to implement.
(28 responses)

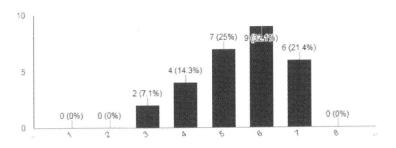

Appendix G

Collaborative contracts with shared incentives for effective risk management lead to better outcomes for Clients.
(28 responses)

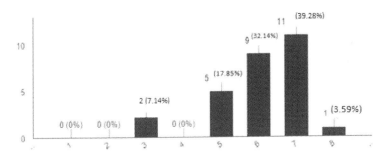

Appendix H

The main risks that are identified at the beginning of projects are usually the ones that come to fruition
(28 responses)

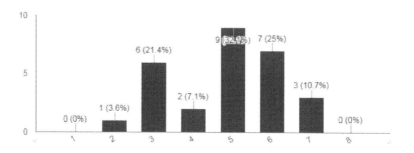

Appendix I

Risk management processes should be an integral part of project management involving the whole team rather than being a standalone process that is managed by one part of the team.
(28 responses)

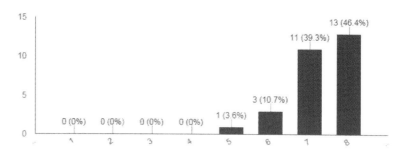

Appendix J

The evaluation of risk and opportunity is influenced more than it should be by the expectations of stakeholders (eg senior managers and /or clients)
(28 responses)

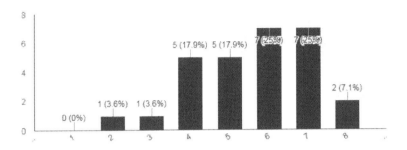

Appendix K

In my experience good contractual arrangements and an appropriate allocation of risk between the parties has a significant effect on whether or not a project is successful
(28 responses)

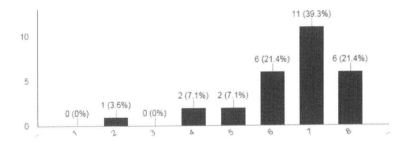

Appendix L

Risk is often passed onto other parties who are lower in the procurement hierarchy even when they don't have the level of influence needed to manage them.
(28 responses)

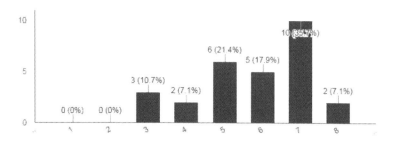

Appendix M

People often avoid releasing unspent risk monies because of the fear of additional unforeseen risks occurring later
(26 responses)

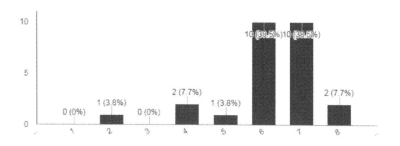

Appendix N

Contractual arrangements sometimes hinder a collaborative approach to mitigating risks
(28 responses)

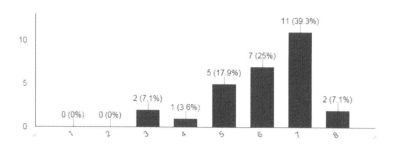

Appendix O

The key to improving risk management in the industry is developing more complex processes to cater for increasing project complexity
(28 responses)

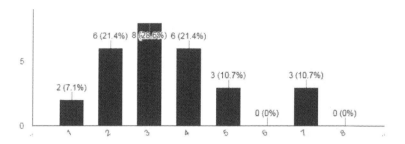

Appendix P

If you have any further comments on how risk management could be improved in the UK construction industry please type below. (if you don't have any further comments, please type 'none')
(14 responses)

Risk management needs to be seen as an overall project management tool and not a stand-alone commercial exercise done to affect the forecast with sensible and achievable mitigation plans and owners

Supply chain need to fully understand they play a role in risk management of any project they are involved in no matter how insignificant it may seem.

Less tick boxes on forms, more discussions in groups.

Key to effective risk management is understanding the blockers early enough to do something them. Effective communication within a team helps to achieve this.

Printed in Great Britain
by Amazon